Variable Frequency Drives: Installation & Troubleshooting!

By

Gary D. Anderson

Practical Guides for the Industrial Technician!

Copyright 2013 Gary D Anderson

ISBN-13: 978-1502770899

ISBN-10: 150277089X

Contents

Chapter 1: Introduction to Motor Control Design   Page 1

- Motor & Control Systems   Page 1
- Why a VFD – A lesson on contrasts   Page 3

Chapter 2: Electrical & Motion Fundamentals   Page 11

- Torque, Speed & Horsepower   Page 11
- Types of Motor Torque   Page 13
- Types of Motor Loads   Page 16
- Electrical & Power Principles   Page 20
- AC Waveform Characteristics   Page 23

Chapter 3: Variable Frequency Drive Fundamentals   Page 27

- Pulse Width Modulation   Page 30
- Carrier Frequency   Page 33
- Fundamental Frequency   Page 34
- Control Modes for Speed & Torque   Page 38

Chapter 4: Drive Programming & Installation   Page 41

- Common Wiring Connections   Page 42
- Parameters & Programming   Page 46
- Menu Navigation & LCD Display   Page 48
- Common Parameters   Page 49
- Braking Methods   Page 57

Chapter 5: Troubleshooting Drive Problems   Page 60

- Basic Troubleshooting   Page 62
- Internal Tests & Checks   Page 68
- VFD Troubleshooting Checklist   Page 77

Chapter 6: Summary   Page 83

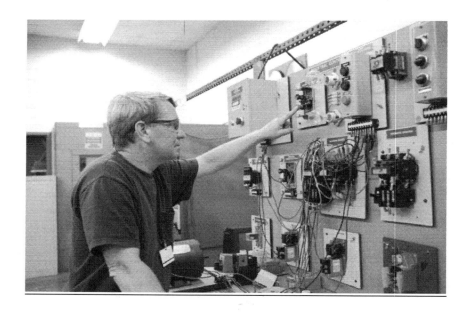

Motor Control has many facets. To fully understand VFD control a solid understanding of traditional control technology is a necessity!

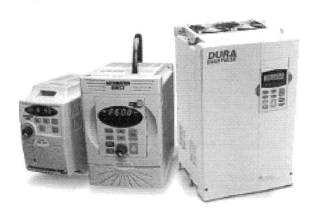

Introduction to Motor Control Design

Motor & Control Systems:

Those of you familiar with modern industry and manufacturing know that there are many different reasons for controlling the speed of an electric motor. These driving factors are usually the manufacturing processes used and also the need to conserve and use energy resources efficiently. So the principle advantage of speed control is to gain the capability to operate at less than full load capacity when it is optimal to do so.

In most any industrial plant, variable frequency drives (VFDs), have become a common and cost effective method of providing speed control and soft-start capability in a wide array of production applications that use AC induction motors.

There are many aspects to consider when designing a Motor/VFD/Control system. In the selection of the necessary components you must carefully consider the application or purpose for the motor, specifically "what will the motor be driving"?

What will the load require from the system, "constant or variable torque, or constant horsepower"? What type of load will the motor/drive system be moving? This may be a conveyor system loaded with coal, a pump moving liquid, a fan moving air, or a CNC machine moving an axis into position.

In addition to these considerations are space requirements, cost of installation and maintenance, and or course the degree of control that must be maintained.

Why a VFD – A lesson on contrasts!

While a VFD may not be the answer in every one of these design issues, I have found them extremely versatile in their applicability in many situations. With advances in solid-state electronics, VFDs stand in stark contrast with older, usually mechanical, methods of dealing with motor speed issues. Remember that older systems still required an electric motor running at full load and speed.

Mechanical measures would then be taken to achieve speed control on the load involved. This might be speed reduction gearing or sheaves, throttle-valves, types of magnetic or eddy-coupled devices or vane-pitch control on a turbine or fan. Since the motor continues to run at full-speed these partial-load methods prove wasteful in terms of energy consumption. Newer Variable-Frequency Drives can often be used to replace older, less efficient systems to reduce energy costs.

Electrical motor control includes traditional full-voltage "*across-the-line*" motor contactors to many other different, and sometimes antiquated - reduced-voltage starting methods. Full-voltage starting, even though being economical to implement, creates much greater demand on the electrical system by drawing up to 700% of the FLA of the motor rating during start-up. Also for loads requiring a lower speed /high torque startup this method is unsuitable.

Other methods of reduced voltage starting would be a wye-delta control set-up, auto-transformer starters, or in some cases a partial-winding type AC induction motor. While providing reduced voltage, lower current start-up, none of these methods provide the capability to operate *continuously* at less than full-load. Below are various diagrams showing these starting methods.

Primary Resistance Starter:

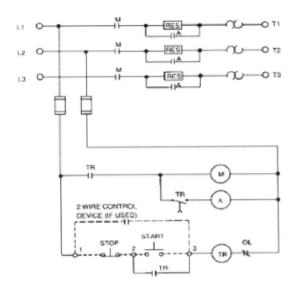

Autotransformer Starter:

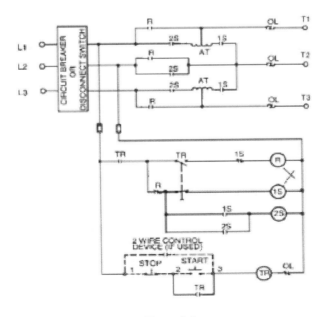

Part-Winding Motor & Starter:

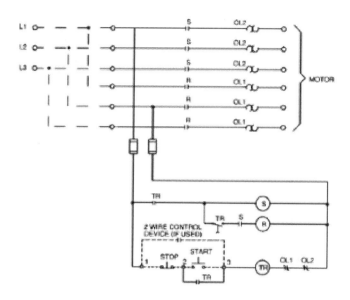

Wye-Delta Starter:

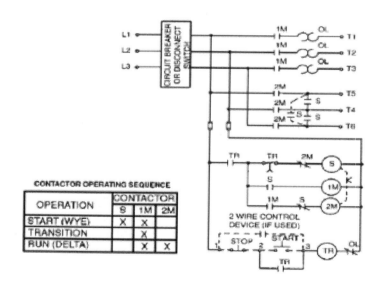

Soft-Start:

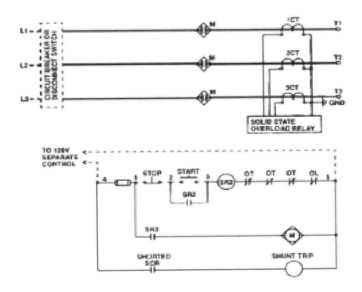

Note that all of these systems, with the possible exception of the "soft-start", use some form of timer to cycle the elements used to provide reduced-voltage to the motor. When the applied voltage is reduced, the motor operates at a reduced speed but also provides less torque. When the motor is operating at a reduced speed there is greater "slip" between stator (the rotating magnetic field) and the rotor of the motor. This increased slip equates to reduced efficiency. Also, when the driven loads still require full torque at reduced speeds these methods cannot be used.

These types of reduced voltage starting mechanisms will on occasion be used to run an AC motor at slower speeds if the application only requires low torque values. However, in most of the above applications, once the timer has completed its function, the control system will apply full voltage and run the motor at its full-load/speed rating. This of course doesn't allow for periods of less than peak demand where it would be advantageous to operate a motor and its load at less than its full-load rating.

The only way for an AC motor to maintain full torque values through the full range of applied voltage is to apply a proportional reduction in the frequency of the applied voltage as well—the *voltage to frequency ratio (V/Hz)* must be constant.

For a motor rated at 460 volts and 60 Hz this ratio would equate to "7.66", (460/60 = 7.66). Thus, in order to control the speed of a AC motor at the greatest efficiency and yet provide full torque value to the applied load, it is necessary to vary **both** the voltage *and* the frequency of the voltage supplied to the motor. In this scenario a "Variable Frequency Drive" or "VFD" controlling an AC induction motor can provide full torque to high inertia loads at lower voltage/current values and can also easily run a continuous duty application at lower speeds if this is required. Here is a simple diagram of a VFD and the various voltage forms throughout each section from its supply input to the output voltage to the motor.

Diagram of typical VFD

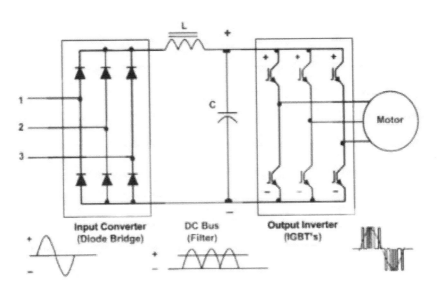

By the 1980s, AC motor drive technology became reliable and inexpensive enough to compete with traditional DC motor control. These variable-frequency drives can accurately control the speed of standard AC induction or synchronous motors. VFDs manipulate the frequency of their output by rectifying an incoming AC current into DC, and then use pulse-width modulation (PWM) to recreate an AC current and voltage output waveform. This will be discussed in greater depth in chapter 3, but for now the above diagram is a simple and effective way to present the changes that occur in the voltage supplied to the motor.

Electrical & Motion Fundamentals

Now that we have covered some of the basic ideas and problems surrounding speed control with common AC induction motors, let's briefly mention some topics that may further clarify these concepts.

Physical Properties: Torque, Speed & Horsepower

Synchronous Speed: The speed of the rotating electrical field in the stator windings of an electric motor. The formula for synchronous speed is:

$$RPM = \frac{120 \times Frequency}{Number\ of\ Poles}$$

Slip: The amount, usually presented on the motor nameplate as a percentage, which represents the ratio difference between the rotor (motor shaft) speed and the "synchronous" speed of the motor. Slip will vary with the load placed upon the motor.

Actual motor speed: The synchronous speed of the motor less *slip*. This is what will be on the motor nameplate.

Power: The rate of doing work. Thus it is measurement of the work done over a specific time period. The common units for describing power are "horsepower" (HP) and watts.

Motors produced in the United States are usually marked with a HP rating on the motor nameplate while in European countries it is more common to see motors with a KW (kilowatt) rating. One HP is equal to 550 lb-ft per second. That is the amount of power needed to lift 550 pounds one foot high in one second of time.

To see how HP and torque are related:

HP = torque x RPM (speed)
Another useful conversion tool is: 1HP = 746 watts.

Torque: It would be quite difficult to understand the fundamentals of AC induction motors without at least a basic understanding of torque. Torque is the instantaneous vector force placed upon a radius, such as a motor shaft. It is basically *rotational force* and is defined in the same terms as

work and energy (lb-ft), or foot-pounds. Torque can be expressed as the following formula: *torque=force x radius*. *As with any force, the torque the motor provides must be greater than the load torque for any work to occur.*
There are four types of torque commonly related to electric motors and their ability to move a designated load.

Types of Motor Torque:

Locked Rotor Torque: The produced torque as full-power is applied to a motor when the shaft is stationary (zero speed).

Pull-up Torque: The produced torque as the motor accelerates to near full rated speed.

Break-down Torque: Basically the maximum torque a motor can produce at full rated speed. At this point the rotor speed is nearly equal to the synchronous speed in the windings of the stator. Beyond this speed the torque of the motor rapidly falls off and can create a stalled motor condition.

Full-Load Torque: The torque produced when a motor operates at its fully rated speed. Remember the concept of the *V/Hz* ratio. If the motor is operating at its fully rated speed and also at "full-load torque", it is because it is operating at its full voltage *and* Hz ratio. This is the value that must be held constant at all lower RPM's for full-load torque to be applied to the load throughout the speed range of the motor.

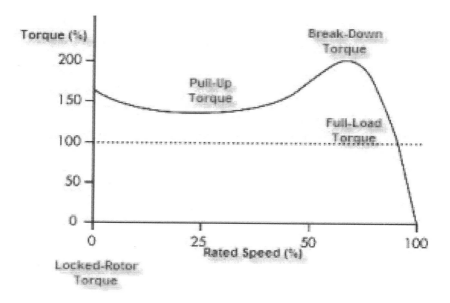

In these diagrams the relationship between torque and HP, frequency and motor speed can be clearly seen. Note the nearly 100% rated speed in the torque curve diagram equates to a frequency of very near 60 Hz on the diagram below.

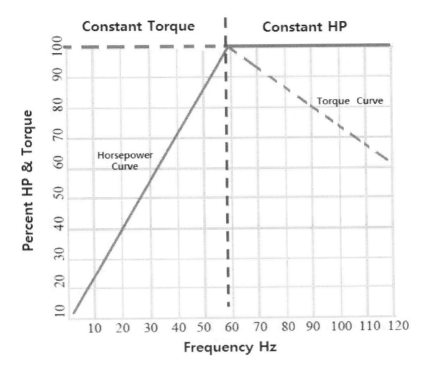

Motor RPM at this 60 Hz frequency range is the speed at which the motor can deliver both constant torque and function very near its fully rated HP. Beyond this frequency level the ability of the motor to provide torque starts to fall off dramatically.

Types of Motor Loads:

To best determine how to drive an AC induction motor it is first necessary to carefully consider the type of load that will be placed upon it. Listed below are the four common load types and common applications for each.

Constant Torque Load:

The load requires that the torque produced by the motor be constant throughout the full speed range of the motor. Torque will remain constant so any increase in speed also increases the horsepower produces at that instantaneous time period.

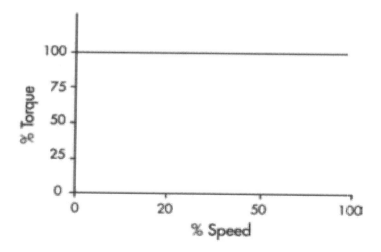

Constant Horsepower Load:

This type of load allows for the torque produced by a motor to decrease as the motor speed increases. Horsepower remains constant throughout the speed range of the motor.

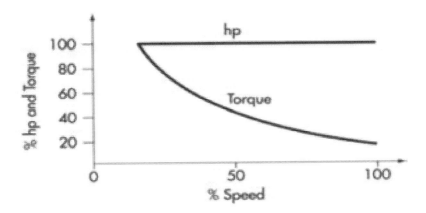

Variable Torque Load:

This is a type of load where the torque required from the motor will increase as its speed increases. For these types of loads torque (t) increases with the square of the speed and horsepower increases with the cube of the speed.

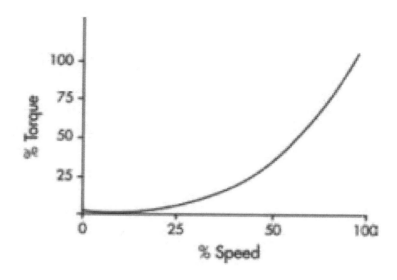

Impact Torque Loads:

A load that may require little from the motor in terms of horsepower and torque to reaching the opposite extreme of values that are several hundred percent of the motor rating.

Summary on Types of Loads:

1. *Variable Torque Load*
   a. Typical Applications:
      i. Centrifugal fans, centrifugal pumps, blowers and HVAC systems.
   b. Characteristics:
      i. HP varies as the cube of the speed.
      ii. Torque varies as the square of the speed

2. *Constant Torque Load*:
   a. Typical Applications:
      i. Mixers, Conveyors, Compressors and printing presses.
   b. Characteristics:
      i. Torque remains the same at all speeds.
      ii. HP varies directly with the speed.

3. *Constant Horsepower Load*:
   a. Typical Applications:
      i. Machine Tools, Lathes, Milling machines, Punch Presses
   b. Characteristics:
      i. Develops the same HP at all speeds.
      ii. Torque varies inversely with the speed.

4. *Impact Loads*:
   a. Typical Applications:
      i. Rock or coal crushers, general high inertia loads
   b. Characteristics:
      i. Wide range of operational load from light to several hundred percent of the motor rating.

Electrical & Power Principles

If you have progressed this far, you have probably noticed that it is assumed you have, at minimum, a basic knowledge of electrical principles and concepts. While it is not the main topic of this text, I would like to mention a few common terms and concepts that are most relevant in understanding VFDs. A solid understanding of these concepts will help both in implementing new VFD systems and also in trouble-shooting existing drive applications

$$\text{Ohm's Law: } \mathbf{\mathit{E=IR}}, \text{ where:}$$
$$E = \text{voltage (EMF)}$$
$$I = \text{amperage (Electron flow or current)}$$
$$R \text{ or } Z = \text{Ohms (Resistance or Impedance to current flow)}$$

The Ohm's Law equation shows how voltage, current and impedance are related in an electrical circuit.

Voltage (E)
The measure of electrical force in the circuit. Conceptually similar to a fluid based system where there must be greater pressure at one point for flow to occur to another point, so too

must there be a differential voltage between points for current flow to occur.

Current (I)

The flow or transfer of electrons through a conductor or circuit and measured in amperage or Amps.

Resistance (R) or Impedance (Z)

The characteristics of a given conductor that oppose or impede the flow of current. The unit of measurement is the Ohm. Impedance is applicative to AC (alternating current), and takes into account factors of *inductive and capacitive reactance* that occur in circuits that utilize alternating current.

When considering impedance or (Z) values it is important to note that it is derived from the resistive load on a circuit and also the inductive and capacitive loads. These last two types of loads present oppositions to current flow in a AC circuit called *inductive reactance* (X_L) and *capacitive reactance* (X_C), both of which are caused by the collapsing magnetic fields intrinsic to AC voltage. Therefore the equation for impedance in a given circuit is as follows and takes into account the total resistance to current flow.

$$Z = R + X_L + X_C$$

AC Waveform Characteristics

Sine Wave

Electrical power in which the values of voltage and current change over a specific time period. This is due to how this type of electrical current is produced – one cycle or wave being one full rotation or 360° of the generator rotor. This is the common way to diagram the voltage and current of AC or alternating current and shows both the positive and negative alternations of current flow.

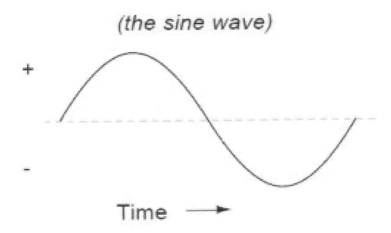

Frequency

This is the number of times a sine-wave repeats during one second of time. The unit of measurement for frequency is the hertz (Hz). It is often convenient to calculate the time period for a single cycle, which of course would be the reciprocal of the frequency. Therefore for a frequency of 60 Hz the *period* would be 1/60 or .0166 seconds.

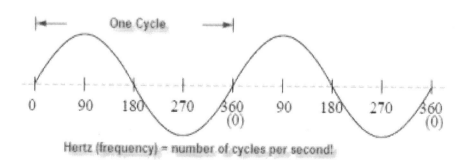

Square Wave

A waveform that rises to specific amplitude, but in which the transition time is very negligible. Basically transitioning from 0 volts to its full voltage in a time period or "rise time" that is zero. Like the sine-wave, the square-wave (opposite page) which is utilized on variable frequency drives, has both positive and negative alternations.

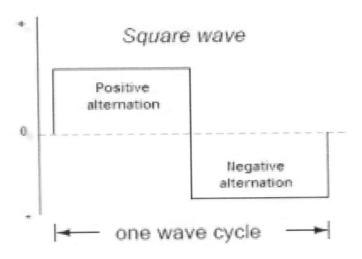

The square wave, as used to create VFD output, is referred to as *pulse width modulation.* The relevant issues in creating the desired output current for the AC motor are:

- *Frequency: How often is the alternation turned to an "on" state - either positive or negative – "off" being zero volts.*
- *Width: How long is the alternation "on".*

Amplitude

The voltage or current presented by a sine-wave during any instantaneous point in time. Various calculations can be made using the amplitude of a voltage sine wave once its peak voltage (PV) is measured. These measurements would include the following: peak-to-peak voltage (PPV), Root Mean Square voltage (RMS), and Average Voltage values.

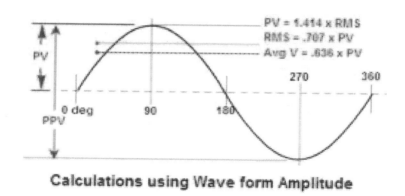

Calculations using Wave form Amplitude

RMS voltage is the mathematical average of the sums of all voltage levels of an AC sine-wave. Therefore the common household voltage of 120 volts, such as in a living room receptacle outlet is derived from a sine-wave that is approximately 170 PV.

Variable Frequency Drive Fundamentals:

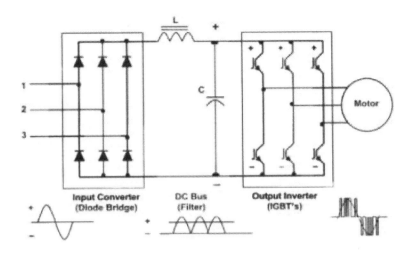

Our above diagram shows the basic circuitry of a typical variable frequency drive that uses PWM (*Pulse Width Modulation*), to provide the desired output current. While there are other types of adjustable speed drives, such as the current source inverter (CSI), and the variable voltage inverter (VVI), the focus of this text is the PWM type which has become an industry standard due to its reliability. The PWM uses IGBT's *(insulated gate bipolar transistors)* in the inverter section which are controlled by a microprocessor and provide very fast switching times.

Notice how the circuit shows three distinct sections. The first section shows the rectifier or *converter* portion of the drive, where a three-phase diode bridge rectifier changes the three phase AC voltage to pulsating DC voltage.

The middle section is the DC bus portion where the pulsating DC produced by the rectification is smoothed to pure DC voltage.

The third section is the *transistor switching* or *inverter* section which produces the three-phase AC current at the desired frequency.

In the first section (rectifier) you can see the six diodes that are connected in a bridge circuit to convert the three phase AC voltage to DC voltage.

The filtered DC section of the circuit consists of several capacitors that are connected in parallel and a large inductor connected in series with the DC bus. The capacitors charge and discharge in synchronization with the alternating input voltage which causes the half wave signal to be converted to a smoothed DC voltage, as denoted by the positive(+) straight line running through the pulses on the diagram. The inductor, labeled by the "L" symbol, is used to filter the current in the resulting voltage. The voltage level at this point in the drive will be at approximately 1.414 times the input voltages. Thus for 480VAC 3 phase input, the DC bus section would be at approximately 678 VDC. The capacitors will charge to the approximate peak voltage of the incoming voltage waveform with little voltage drop through the diode bridge.

The transistor or "inverter" section of the VFD consists of our six IGBT's. You can see that one transistor of each phase is connected to the positive DC bus and the second transistor is connected to the negative DC bus. These IGBTs are switched to "on" and "off" states by a firing circuit controlled by a microprocessor located within the VFD. At the correct time each transistor is turned on which applies a voltage pulse to the output of the drive and to the AC motor. The following example shows the resulting waveform that is applied to each motor lead or phase. The action of the inverter section takes the DC bus voltage, and using Pulse Width Modulation (PWM) sends an applied voltage to the motor which appears as AC current. _Both the voltage amplitude of this hybrid output and its frequency are functions of the length of time the voltage pulse is turned on and the time between pulses._

Pulse Width Modulation

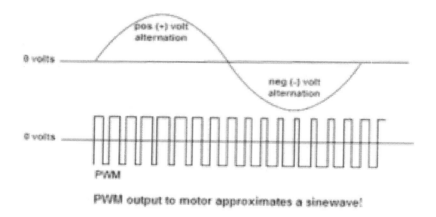

PWM output to motor approximates a sinewave!

In other words, the motor is seeing the applied voltage as a sine wave, which is created by the VFD's controlled pulsing of the filtered DC voltage in precise time sequencing. The longer the pulse is "on" – the higher the resulting voltage output. Please recall that an AC voltage is the average of the peak voltages of each alternating half-cycle. You can see from the above diagram that the pulses that have the longer "on-time" coincide with the higher amplitude in the resulting sine wave.

Example of Pulse Width Modulation

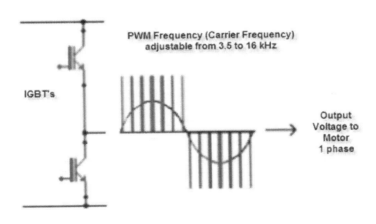

The output frequency is adjustable from 0 to 60 Hz with the *V/Hz ratio* remaining constant. In other words for every decrease in frequency the drive will also lower the output voltage. This holds for frequency selections up to 60 Hz. Above this frequency value, the output voltages will remain constant and the motor will lose efficiency. The following diagram (shown earlier) visually depicts how both torque and HP are related to frequency.

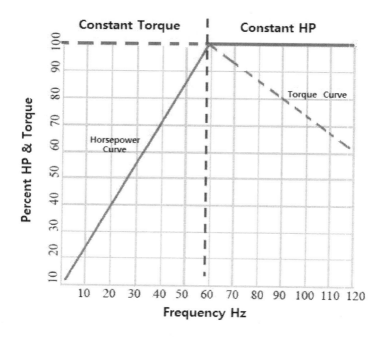

Carrier Frequency:

Carrier frequency, often referred to as *"switching frequency"* is the term used to denote the frequency of the square-wave pulses produced by the inverter section of the drive. The firings of the transistors are determined by the microprocessor and can usually be set from 3 kHz to 16 kHz. This setting will determine how often the drive sends the pulse groups to the motor. As the setting on the switching frequency is set higher, the resulting current waveform is tighter in resolution or "smoothed". However, carrier frequencies less than 3 kHz are audible (and unpleasant) to the human ear, therefore carrier frequency is commonly set within the 4 to 8 kHz range, and sometimes higher.

At these frequencies the carrier frequency closely approximates a pure sine wave. The more closely the current delivered to the motor resembles a pure sine wave, the cooler the motor will run.

Fundamental Frequency:

This is the frequency of the voltage used to control the speed of the motor. Usually between 0 and 60 Hz, but can be set to go above the full-speed (60 Hz) rating of the motor.

To create a fundamental frequency of 60 Hz current to a motor, with the carrier frequency set at the 4 kHz setting, the microprocessor would fire the IGBTs in such a way as to cause them to send 66 pulses of the DC bus voltage to the motor lead for every 60 Hz cycle. So the VFD would send 33 pulses for the positive alternation (half-cycle) and 33 for the negative alternation. Remember that the period for a 4 kHz frequency is .00025 seconds and the period for our desired 60 Hz output is .01666 sec.

.01666 / .00025 = 66.64 total per output cycle (60Hz)
Therefore: 33 per half-cycle

At 8 kHz the number of pulses would double to 133, assuming a full-speed output frequency of 60 Hz is desired, creating a smoother current sine wave to the motor.

Note that as carrier frequency is set higher, it can also produce voltage spikes that damage insulation on wiring and on the motor itself. This has become less of a problem if motors are wound with inverter rated wiring and also if distances between the VFD and motor are kept to within 100 ft. At longer distances, inductors or chokes are often used on the output of a drive to filter these voltage spikes and prevent damage to wiring and equipment.

Before moving on to the important topics of programming, setting parameters and the topic of troubleshooting, carefully study the following diagram that is a simple analysis of the PWM waveform. When using an oscilloscope to view output voltage, the pulse trace would of course be what is seen.

Using *current* settings on the oscilloscope, the output sine wave could be viewed. The motor would see this resultant voltage/frequency as the current sine wave that is superimposed over the pulse train.

PWM sine-wave analysis:

Smaller pulse widths produce a lower output voltage at a "longer wavelength" resulting in a lower frequency. *V / Hz ratio remains constant!*

Wider pulse widths produce a higher output voltage at a "shorter wavelength" resulting in a higher frequency. *The V/Hz ratio remain constant!*

Pulse Width

Current Waveform to Motor!

DC Bus Amplitude
Voltage for 480 VAC input:
480 x 1.414 = 650 Vdc

Pulse Width

One Cycle

Methods of Speed and Torque Control:

So far we have discussed the whole idea of maintaining a constant V/Hz ratio to maintain full torque throughout the speed range of an AC induction motor. Due to advancements in solid state technology there are now _several methods of achieving this goal_. Each one of these _control modes_ offer differing characteristics and benefits in terms of speed control and _torque bandwidth_, which is the optimum range of torque values a drive can provide given a specific frequency.

Maintain a constant voltage to frequency ratio by:

Scalar control: Maintains a fixed V/Hz ratio over its operating range. Once established by the control set-up procedure the voltage supplied to the motor at various operating frequencies is determined and controlled by this ratio. For a 460V/ 60Hz motor the ratio would be 7.67 and for a 230V/60Hz motor the ratio would be 3.83. These ratios are maintained unless voltage boost or IR compensation is activated, or the frequency is increased beyond the level for which the system can maintain the proportional voltage. Ramp time adjustments are used to prevent acceleration and deceleration currents from exceeding safe limits. This method provides a torque bandwidth of approximately 10 to 50 Hz.

Open-loop flux vector control: Mathematically estimates speed to control the flux and torque producing currents in an AC induction motor. This type of control is also referred to as _sensorless vector_ control. Vector drives dynamically regulate motor torque as directly and accurately as possible by continuously monitoring and analyzing the motor current to determine what voltage to apply at any given frequency to produce the _optimum magnetic flux_ in the motor windings. This method can provide torque bandwidth of .6 to 300 hertz.

Closed-loop flux vector control: Directly measuring speed, usually by means of an encoder, to control the flux and torque producing currents to an AC induction motor. This type of control is often referred to as _vector control_ and can provide a wide range of torque bandwidth from zero speed to 500 Hz. Vector drives, including the "open-loop" or "sensor-less" type can often provide more than 150% of the rated torque to smoothly accelerate the load.

Direct torque control: in which the microprocessor uses two control loops – a speed control feedback such as an encoder and also a torque control feedback loop.

In this manner the processor is able to monitor both the actual speed and also the torque load on the system and can perform inverter switching that will satisfy the calculated "error". This correction happens in microseconds making this a very efficient type of control mode for many applications. Speed accuracy for this type of control is in the range of 0.1 to 0.5 percent of the slip rating of the motor (nameplate rating).

Drive Installation & Programming:

As with any type of electronics the best installations are those where equipment can be kept clean, dry and reasonably cool. Most VFD's have an operating temperature range of 0° – 40° C, or 32° to 104° F. Most VFD's will also have a cooling fan or fans integrated into the drive to provide airflow cooling over the heat sink. With any installation, ventilation and access for maintenance should be a main consideration, it makes troubleshooting all the more difficult when a technician must take test measurements or make adjustments in a poorly lit, difficult access area. If your drive is rated and setup properly, in all likelihood it will not require much attention for many years, but best practices would mandate good accessibility. Some common issues with VFD drives may be the need to periodically clean filter media in control cabinets, replace cooling fans, capacitors or diodes, make minor programming or parameter changes, and the list goes on. So, - when you install a drive – make it as maintenance friendly as possible, you will make it easier on yourself and others as well!

Common Wiring Connections:

Aside from main input power for the drive, -normally some form of 3-phase power, and the motor output terminals, a VFD will also employ a number of terminal connections. These terminals and their associated wiring will control when and how the drive operates, and many are configurable by parameter options. Drives usually have terminals for digital inputs that allow for 2-wire or 3-wire control, start, stop, reset, forward and reverse, a run/enable and jog input and also some digital inputs for preset (by parameter) speeds.

While it is convenient to show 2 and 3 wire start/stop wiring with conventional pushbuttons, remember that these digital inputs can be (and often are) initiated from a programmable control platform such as a PLC or a CNC machine control processor.

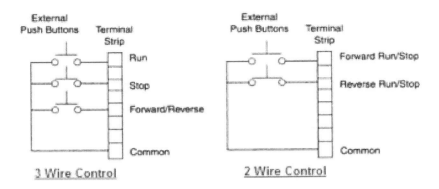

Normally drives will have the option of using an analog input, such as 0 to 10 vdc or 4 to 20 ma, to control its output. In many applications the control wiring for the drive is via Ethernet cable and networked from an OEM control platform such as a PLC. In these scenarios the only additional wiring that might be present in the drive is perhaps wiring for the braking resisters.

Here is a diagram that shows some of the terminal connections that seem relatively common to all drives. It is important to note that not all of the connections shown are necessary for the drive to run properly and that many are dependent upon parameter setting.

Toshiba VF-S11

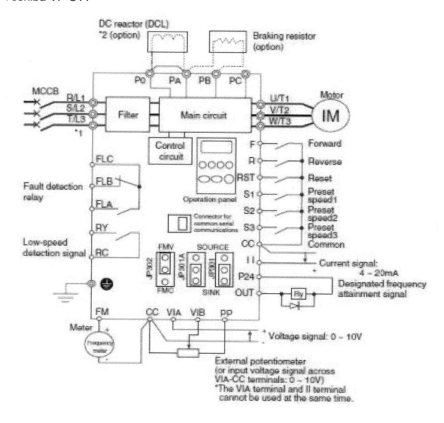

Usually the voltage at control terminals is derived from the drive itself, so no additional power beyond the incoming 3-phase voltage will be necessary. Speed adjustment will be accomplished by external potentiometer, 4-20 ma current signal, or perhaps by a 0-10 vdc analog input. These selections are normally set by parameter but may also be set by a jumper or dip switches on the control board. If a pre-set speed is utilized from one of the digital inputs, note that the input is looking for either a "high" or "low" logic state, the actual speed/frequency associated with that input is a value that will be set by specific parameter.

Fault detection contacts that are integral to the drive (see FLB above), will be used to shut down the drive and also other related equipment in the event of a drive fault or failure. CNC machinery, which often utilizes multiple drives for multi-axis machining will usually connect these contacts in series or "daisy-chain" all the drives together in this fashion. If any single drive fails – the machine goes into an e-stop or fault condition.

Parameters & Programming

Programming is one of the most important aspects of installing and maintaining a VFD. It is accomplished by the setting of different parameter values or by the selection of certain "macro" programs within the drive memory. These macro programs can greatly speed up the setup process and are also selected by specific parameter settings.

Parameter setting is most often accomplished by some form of keypad display interface that is integral to the drive. In some models, a drive will utilize a "personality module" which must be programmed with a PC using specific application software for that drive. In either case, programming is done by going through the parameter groups and setting desired values. Each parameter will specify a property that will determine certain operational characteristics of the drive. *What type of control mode* - Constant V/Hz, Sensorless Vector, Closed-loop Vector or Direct Torque? *What type of* _speed reference_ - Pre-sets or some form of analog signal? _Will additional output contacts be used and in what way_? All of these questions and many more must be decided as you setup or program a drive to perform as you desire.

Many manufacturers include hundreds of parameters that can be used. Even a relatively small drive will often have multiple pages of parameters listed in their manuals. While I usually scroll through every one when I do a drive setup, I have also found that many times only a small portion of these, maybe 20 or so, are necessary.

It should also be noted that drives are often shipped "pre-programmed" with default parameters. These generally are set to conservative and frequently used values that will pose the least risk to equipment and personnel.

Menu Navigation / LCD Programming Display

On most programming displays the software will be arranged in a menu-based system. Parameters will usually be assigned to a specific group and also functional sub-categories.

One group will define the characteristics of the drive such as what language it will display, time and date settings, and determine what it will display on the display panel, such as the frequency, voltage, current or motor speed (in rpm). Another group will be the specific information about the motor being used – much of which will be taken from the motor nameplate. Other groups will contain the information that determines exactly how the drive will operate, what frequency will it provide, what frequency will be "skipped" over, what is the switching frequency desired, etc. One important aspect of the keypad/display unit is that it will usually allow the technician to bypass the remotely wired controls and to run the drive directly from the keypad (local mode).

Common Parameters

The following table is a summary of 19 commonly used parameters and some description of each. We will end this section with a brief discussion of "braking" and then close with the section on "troubleshooting".

P#	Description	Default	Range
P-00	Remote Enable	OFF	ON / OFF
P-01	Acceleration Rate (sec)	5.0	0.5-30.0
P-02	Deceleration Rate (sec)	5.0	1.0-30.0
P-03	Minimum Speed (Hz)	5	3-30
P-04	Maximum Speed (Hz)	60	30-140
P-05	Current Limit (%)	150%	10%-150%
P-06	Manual Torque Boost (%)	2%	0%-10%
P-07	Volts/Hz Base Speed	60	30-240
P-08	RPM at Base Speed	1750	1-9999
P-09	Output Relay Enable	OFF	ON / OFF
P-10	Carrier Frequency (kHz)	8	4,6,8,12
P-11	Remote Reference Gain (%)	100%	60%-100%
P-12	Remote Reference Offset (%)	0%	0%-40%
P-13	Remote Reference Display Enable	OFF	ON / OFF
P-14	Electronic Thermal Overload (%)	100%	20%-100%
P-15	Thermal Overload Enable	OFF	ON / OFF
P-16	Coast Stop Enable	ON	ON / OFF
P-17	Reverse Disable	OFF	ON / OFF
P-18	Software Version	Read Only	N/A

Remote Enable: This setting will allow the drive's starting and stopping options to be controlled either in "local" mode via the drive keypad or in "remote" mode which would be from remotely located switches or control wiring.

Acceleration Rate: The "accel" time is the time period, usually in seconds, that the drive will take to accelerate to its full or programmed speed. This is basically a "soft-start" feature that limits inrush current to the motor and allows a smooth startup of load movement. A proper setting with this parameter will prevent "overcurrent faults" on the drive.

Deceleration Rate: The time period, usually in seconds, that the drive will take to decelerate the motor from full or programmed speed to a full stop. It is important to note that this setting is very dependent upon the load being driven. If it is a high inertia load with large mass then decel time must be adequate to stop the load. If too short a time period is set then "overvoltage" faults will occur on the drive, a DC bus fault, because of regenerated energy pumped back into the

VFD. A braking resistor grid (dynamic braking) is often used to dissipate this energy.

Minimum Speed (Hz): A minimum frequency setting for output to the motor that is valid regardless of any other regulated speed setting from an external control.

Maximum Speed (Hz): A maximum frequency setting for output to the motor that is valid regardless of any other regulated speed setting from an external control.

Current Limit (%): This parameter allows the user to set the maximum percent of the drives current rating. This parameter setting will limit output torque and also the set accel and decel times to keep the drive within its current limit setting.

Manual Torque Boost (%): This parameter is sometimes referred to as "IR Comp". It is set as a percentage, and will provide additional voltage at very low starting speeds. This will cause the motor to provide extra torque during the startup of high inertia or high friction loads. This setting can cause "overcurrent" faults if set to high.

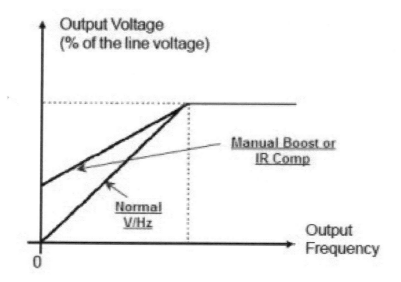

Volts/Hertz Base Speed: This parameter sets the base speed at the maximum output voltage and establishes the V/Hz ratio that the drive will follow through all speed or frequency ranges up to this maximum value.

RPM (Base Speed): This parameter scales the displayed RPM, if this option is used, to the base (nameplate rated) speed of the motor.

Output Relay (Configurable): This parameter selected how the integrated output relay is energized for the purpose of showing different drive conditions such as "running", "faulted", or "at speed".

Carrier Frequency (kHz): With this parameter the user can select the frequency with which pulses are sent to the motor output. In the above table the selections are 4, 6, 8 or 12 kHz but many drives are selectable to 16 kHz. Remember that lower carrier frequency settings, such as 2 kHz – 3 kHz, can cause the drive to produce audible noise which can be irritating. The higher the carrier frequency, the smoother the current signal will be to the motor and the smoother the motor will run. The downside of higher switching frequencies is increased heat in the inverter section due to faster switching times on the transistors and also the potential damage to motor insulation due to voltage spiking.

Remote Reference Gain (%): This parameter allows the user to scale the actual analog speed reference to a maximum other than the 10 vdc (if using 0-10vdc) or 20 ma (if using 4-20ma). For instance, if max speed needs to be referenced by 16ma rather than 20ma,, then the *Reference Gain* would be set at 80%.

Remote Reference Offset (%): Similar to the Reference Gain, this parameter scales the minimum analog signal to the actual range of the reference signal. For instance if using a 4-20ma signal, the minimum would not be zero or 0 ma, it would be 4ma. Therefore the *Remote Reference Offset* would be 4/20 or 20%.

Remote Reference Display Enable: Allows the user to turn "on" or "off" the displayed speed.

Electronic Thermal Overload (%): This parameter selects the trip setting for the motor overload fault. Remember that the motor current rating may be less than the VFD current rating. Divide the motor current rating by the drive rating for the percentage setting.

Electronic Thermal Overload Enable: This parameter, if set to "ON", enables the thermal overload function and will protect the motor from overload to the value selected by the Electronic Thermal Overload (%).

Coast Stop Enable: If this parameter is enabled the IGBT's will simply turn off when given a stop command, which will then allow the motor to coast to a stop. Otherwise, if disabled, the IGBT's will continue to provide systematic firing to ramp the motor down and then turn off.

Reverse Disable: If set this parameter prevents the motor from being driven in reverse by the VFD.

Software Version: Usually a read-only parameter. It is important, when troubleshooting a drive, to have all the details, reference manuals, and also the software version before calling the manufacturer for technical assistance.

S-Curve Parameter: Not listed in the above table but often used to provide additional smoothing or "soft-start" and "soft-stop" capabilities. It makes adjustments to *not allow* a purely linear start or stop. This is a "time-based" parameter.

Critical Frequency or Skip Frequency: This parameter setting will allow the drive to "pass-over" certain fundamental frequencies that trigger mechanical resonance or harmful vibrations in equipment.

Automatic Restart: This parameter, for obvious safety reasons, usually has a default value of "OFF". It can, when the application allows, be set to "ON", where the drive will automatically restart if a non-critical fault condition has cleared.

Braking Methods

These methods must be considered whenever installing or configuring a new adjustable speed drive to an existing application. The method you use will be dependent upon such factors as mass, inertia, and friction. These basic methods of stopping a load are set by parameter but may, as in the cases of regenerative braking or dynamic braking also require additional equipment and drive capabilities.

- Coast to a Stop
- S-Curve Stop
- Ramp to Stop
- DC injection Stop
- Dynamic Braking Stop
- Regenerative Braking Stop

Dynamic braking requires a resistor grid to dissipate (in the form of heat) the electrical energy that is forced back to the drive.

A drive with *regenerative braking* utilizes forward and reverse IGBT bridges that, when fired will dissipate electrical energy back onto the mains.

Here is a comparison chart for the relative stopping times using different forms of braking.

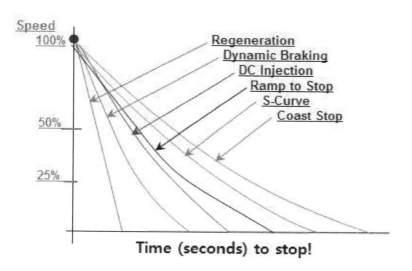

Once again, the method you choose will depend on the load, how quickly it must come to a complete stop, and also whether the costs of additional equipment, such as braking resistors or a drive with regenerative braking, is acceptable.

Troubleshooting Drive Problems:

Many technicians know the feeling you have when production equipment goes down due to some mysterious unknown, operators are standing around with hands-in-pockets, and production managers are asking – *"how long will it be down?"* or saying *"we need it going now!"*.

All of the previous chapters have brought us to this point and hopefully will have prepared you to quickly and safely isolate problems with VFD driven equipment. What I have attempted in this short book is to present concepts that are absolutely necessary to the successful troubleshooting of variable frequency drives. There are many other areas of VFD utilization that could be discussed, such as motor classes, types and enclosures, formulas regarding power factor and harmonics, and detailed cost-analysis of using these drives. All of these things, while important, have not been discussed because I wanted to present a basic framework of concepts that would bring about good understanding and to aid you, the electrical technician, in troubleshooting and problem resolution. With that goal in mind we continue with basic and necessary troubleshooting concepts that will – almost 100% of the time help you isolate and resolve drive problems.

Variable frequency drives (VFDs), consist of a complex combination of electrical power components and sensitive electronic circuits. They are controlled by various feedback and microprocessor circuitry that can be very sensitive to induced voltages from other sources. Another important thing to remember as you troubleshoot a problem is that proper VFD operation is dependent on many external factors such as the nature of the load, input voltages, power surges, proper cooling, moisture, ventilation and cleanliness.

If you want to be effective in troubleshooting VFDs, or any other electrical problem, you must apply a systematic and common-sense approach to your troubleshooting techniques.

Basic Troubleshooting

As in all electrical problems, there are basic procedures and methods to always perform first when troubleshooting a VFD application. *Safety should always be your major consideration and the first step in good troubleshooting practices.*

Let's repeat that:

Safety should always be your major consideration and the first step in good troubleshooting practices.

Be sure to install warning signs, place barriers around the test location to prevent pedestrian walk-thru traffic, and follow proper lock-out procedures.

Take as many test readings as possible with the *power off*. However, many measurements and test procedures will have to be done on energized equipment. Important readings to take are: *Input voltage, Output voltage, Output current, frequency reading on the output, DC bus voltage*, and a *resistance check on motor leads*.

It's important to remember that both AC and DC voltages are present within a VFD. As previously discussed, DC bus voltage can be at very high and dangerous levels. It will be equal to the peak voltages of the AC input, approximately 680 vdc, and this charge will be held by capacitors that are parallel across the bus. A potentially lethal charge can be held for several minutes before dissipating so always take a voltage reading, with a verified meter, on drive capacitors or DC bus before performing any hands-on work.

<u>Is power present to the VFD input terminals</u>? <u>Is the breaker tripped or fuses blown?</u> <u>Might the DC bus fuse be blown?</u> One of the most important questions of all, <u>what does the machine operator have to say?</u> They often give important details about the situation at hand. Possibly a fault occurred as they were making tension or speed adjustments?

When these questions have been answered, then go to the drive to see what fault indications may be showing. Included here is a list of common drive faults and some of the issues that cause them. These will many times be displayed on the drive keypad/programming module but it is still necessary to have the manufacturer's manual with you. Some common drive faults are:

Over temperature fault: This can occur if the cooling fans that circulate air over the heat-sink have stopped working, or if the heat-sink or the filters are clogged and dirty. On occasion the thermistor is simply bad or has a faulty connection.

Over current fault: This fault will occur if the "overload" parameter is incorrectly set for the size of the motor or if the rating of the drive is simply inadequate for the motor it must drive. Also remember that if extremely high carrier frequencies are used that the drive may need to be "de-rated" which could make it unable to produce the current necessary for the application.

Of course, if the driven load is jammed or a seized motor could cause this fault as well. If one phase of incoming power is lost to the drive it will continue to provide 3 phase output to the motor but at substantial power loss.

Over voltage fault: This fault can occur when incoming voltage to the drive is too high, so it is important to check and verify incoming power. Basically this is a fault that concerns the DC bus voltage. If the incoming power is high, the bus will carry the approximate PV of the incoming AC phases.
Usually, when I have encountered this fault, it is because of either a "deceleration" parameter being set at too short a time period, a heavy load is "overhauling" the drive, or the regenerative braking is not set properly.

These faults can usually be reset from the keypad or by turning off power to the drive and then back on - but remember to look for *root causes* in these scenarios.
Another good practice is to check the integrity of the motor (or motors) connected to the drive. This can usually be

accomplished with a DVM and a "megger". Megger or "*ground-insulation*" testing is usually done at the 500 Vdc or 1000 Vdc settings. As a general rule, a motor in relatively good condition will give a reading of 50 meg-ohms or greater to ground potential. Also, using a megger can test the integrity of the load wiring that feeds the motor. Any low readings can indicate the breakdown of insulation either in the motor or other load wiring. Of course this is often caused by environmental conditions such as oil, coolants, moisture and dirt.

If motor and wiring checks are good (or within reasonable limits) you can remove motor leads and simply see if the drive will run without any load. If the motor tests "good" then you can select the "local" option on the drive keypad and see if you can drive the motor in that fashion. If so, then the problem resides in the control wiring "external" to the drive. Has the drive lost an analog speed reference or the enable/run signal that it requires to operate? If connected to a control platform, such as a PLC, are the wiring and cable connections good?

If the drive was producing erratic speed control then the speed reference signal wiring may be acquiring background noise. Is it shielded and grounded properly and are the connections good? Is feedback wiring routed near power wiring in the machine controls cabinet? If so, this can be the cause of erratic movement due to unwanted noise on the signal control or feedback wiring.

If these issues have been addressed then it may require some tests to check and verify the internal components of the VFD. At the end of this section I've included a check-sheet that can be used to check many of these internal components in a systematic method.

The VFD: Internal Testing

Often, drives do not have their own integral power supply but share a common "machine power" AC source and a common DC power supply and bus. This can make it a bit easier to check the DC voltage but remember that this voltage can be quite high – over 600 volts if the input voltage is 460 VAC, so great care must be taken when troubleshooting a drive. At the other end of the spectrum are the drives in which all the sections, "*converter-DC bus & filtering-and inverter*" are contained in one package. The drive, like any other electronic component, can be checked and verified, section by section and component by component if necessary. Usually, as in the examples below the DC bus will have terminals that are accessible and can be used for test purposes. Note that some drives have very good "onboard" diagnostics which may tell you if a processor, base-driver firing board, or output transistor is at fault. It is usually necessary to consult the product maintenance manual for decoding the many "fault codes" that comprise these diagnostic aids.

I recall a large spindle drive on a 5-axis milling machine that had a row of numbered test pins that could be checked for a high logic state – 5 vdc or so. This row of test pins, depending on which were high and low state, provided the onboard diagnostics that could be interpreted (using the manual), to isolate different problems within the drive. Most VFDs today will use some form of digital display to present an alarm condition or fault code.

Next are some examples of DVM tests; I use a Fluke 87 meter and, for this example a Toshiba VF-S11 drive. Test points for input line (L1, L2, and L3), output (T1, T2, T3) and the DC bus (PA+, and PC-) are all brought to the front terminals which made this an "easy" example to use. Please note that these diagrams show a power transistor with a protective diode on the inverter output side. As previously discussed, today it is much more common to find IGBTs *(Insulated Gate Bipolar Transistors)* in use. They have very fast switching times and use less current on the control side than a conventional power transistor, while providing for high current output. IGBTs are basically a hybrid between an IGFET and a normal power transistor.

Test of the Upper Diodes on Rectifier (converter) Section

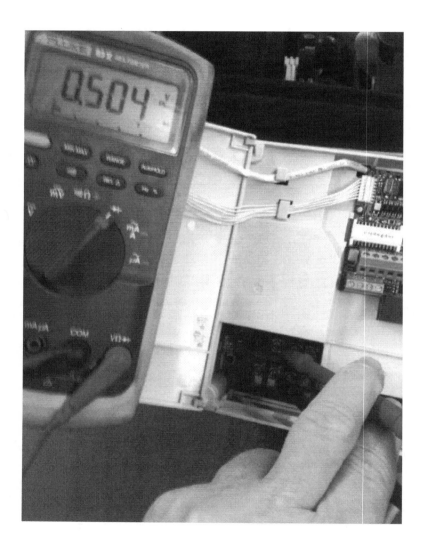

Test each input terminal in this configuration; readings should measure a .3 vdc to .6 vdc voltage drop using the *diode check* selection on your meter. The difference depends on what type of solid state components, "germanium" or "silicon" and the "doping" procedure that turned their junctions into a "P-type" or "N-type" solid-state device. If your readings are open or shorted then the diode is bad.

If this test is good, then reverse the leads – negative lead to the input terminal and positive meter lead to the DC+ bus, (PA+) in my example. You will probably see a brief charge-up of the capacitors but the meter reading should go to "OL" or read like an open showing no conductance. If readings are shorted then the diodes are bad.

This same method can be used to test the lower diodes in the bridge and the results should be the same as on the other 3 diodes.

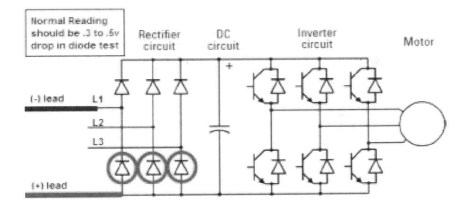

At this point, it may be convenient to test the capacitors. One simple test is to put your meter leads across the bus terminals PA+ and PC-, while in the *diode test* mode. You can see the caps charge to the voltage your meter can output, probably 6 vdc or so. As you switch the selector on the meter to read DC voltage you will read the charged voltage and see it slowly dissipate. Of course, a better test is made with an oscilloscope while the drive is energized to see if the voltage is steady and without dips or AC "ripple". Also it is a good idea to visually inspect capacitors for swelling, leakage and physical damage.

Next we can test the output (inverter) section *protection* diodes. If these have failed then in all likelihood the transistor associated with that protective diode has failed as well.

Tests for upper and lower output protection diodes:

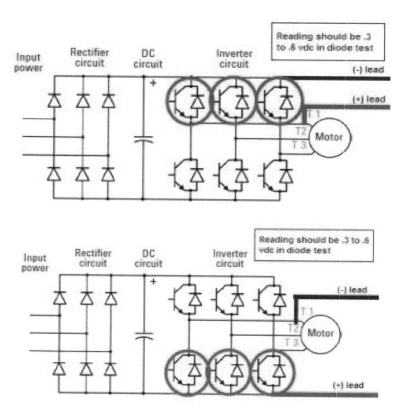

Once again, note the placement of test leads and check for shorted and open diodes. The inverter transistors can be difficult to test because of requiring removal but a quick check between output terminals, while in diode test mode, should show no conductance or "OL". Older style, "Darlington pair" transistor packs could be tested but I always found it convenient to have the "datasheet" handy when testing these units.

If these tests are good, then other potential problem areas may reside within the control and microprocessor area of the drive, the internal relays of the drive, or the internal switching power supplies of the VFD. All of these are areas in which it may be more cost effective to purchase a new unit depending on the size and HP of the drive. Many times a power supply can be acquired for replacement or repaired just like any of the other components in the drive package, but the cost in lost production may be the deciding factor.

These are the general procedures for isolating a problem on a VFD. These same guidelines seem to work well regardless of the AC drive or its application. PWM drives are common in CNC machine tools with AC servo motors and also a wide array of pumps, fans, lift tables, and conveyor systems.

Included at the end of this section is a checklist that can be used as a reference when doing these internal tests.

VFD Internal Checklist

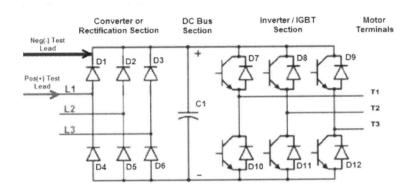

1. Measure Rectifier Forward Vdrop (diode test setting)
   Normal readings: .3 to .5 volts, Reversing test leads should read OL.

   A. Red Lead to L1, Blk Lead to + DC bus:
      D1FVdrop _____
   B. Red Lead to L2, Blk Lead to + DC bus:
      D2FVdrop _____
   C. Red Lead to L3, Blk Lead to + DC bus:
      D3FVdrop _____
   D. Red Lead to – DC bus, Blk Lead to L1:
      D4FVdrop _____
   E. Red Lead to – DC bus, Blk Lead to L2:
      D5FVdrop _____
   F. Red Lead to – DC bus, Blk Lead to L3:
      D6FVdrop _____

2. IGBT Freewheel Diode Forward Vdrop (diode test setting) Normal readings .3 to .5 volts, Reversing test leads should read OL.

   A. Red Lead T1, Blk Lead to DC+ bus:
      D7FVdrop: _____
   B. Red Lead T2, Blk Lead to DC+ bus:
      D8FVdrop: _____
   C. Red Lead T3, Blk Lead to DC+ bus:
      D9FVdrop: _____
   D. Red Lead DC_{neg} bus, Blk Lead to T1:
      D10FVdrop _____
   E. Red Lead DC_{neg} bus, Blk Lead to T2:
      D11FVdrop: _____
   F. Red Lead DC_{neg} bus, Blk Lead to T3:
      D12FVdrop: _____

3. Bus Capacitor Series Resistance and Capacitance:

   A. Ohm scale: Red Lead to (+) and Blk Lead to (-) (if resistors are parallel across the Bus capacitors):
      _____ ohms

   B. Capacitor test mode: Red Lead to (+) and Blk Lead to (-): _____ microfarads

   C. Remove Capacitors and inspect for swelling, leakage, or other damage.

   D. Test each individual capacitor for proper readings:
      _____ microfarads

4. Digital multi-meter (DMM) test procedure for IGBT modules:

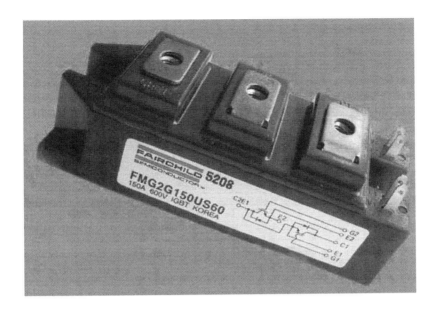

IGBT Schematic Diagram

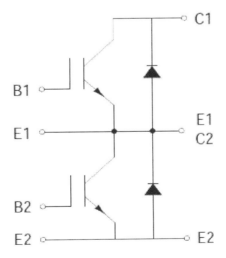

Collector – Emitter Junction Test

(B1 & B2 may be labeled G1 and G2 on the IGBT)

1. Short out G1 to E1 and G2 to E2. With a DMM set to diode test mode, check across the C1 - C2E1 junction. With the (+) probe on C1 and the (-) probe on C2E1, you should see an open circuit. Switch the probes. You should see a diode drop on the meter.

2. Check across the C2E1-E2 junction. With the (+) probe on C2E1 and the (-) probe on E2, you should see an open circuit. Switch the probes. You should see a diode drop on the meter.

3. With a 9 volt battery, connect the (+) terminal to G1 and the (-) terminal to E1. Using your meter (set on diode test), you should now see a diode voltage drop across the C1-C2E1 junction in both directions now.

4. Connect the battery (+) to G2 and the (-) to E2. You should see a diode voltage drop across the C2E1 - E2 junction in both directions here as well.

5. If the IGBT passed all of the above tests, the IGBT is good.

Gate Test:

1. With the DMM set to test resistance, check between the gate to collector and gate to emitter. These should read infinite on a good device. A damaged device may be shorted or show leakage from gate to collector and/or emitter.

Notes:

Notes:

Summary

I can say that most of the problems or "fault" scenarios I have encountered with VFD's have, in the final analysis, been issues with the application of use rather than actual drive failures. These very reliable devices tend to do, or try to do, exactly what we program and ask of them. If we accidentally program a maximum speed that is twice the rated speed of the motor, the drive will attempt to provide that speed – whether it damages the equipment or not, - at least for a short period of time.

While it is often tedious to read a manual, it is extremely important to be thoroughly familiar with the drives you use, their applications, programming procedures and how to perform safe lock-out / tag-out procedures as well.

I hope this guidebook has been helpful in developing a greater understanding of "adjustable speed drives" and their importance in industry. Hopefully, it will also be beneficial to those of you who are often called upon to service and maintain complex equipment in, what are sometimes difficult and demanding situations.

***

Practical Guides for the Industrial Technician!

Motion Control Basics: Troubleshooting Skills for CNC & Robotics!

Variable Frequency Drives: Installation & Troubleshooting!

Industrial Network Basics!

Note from the Author:

Thank you for taking the time to purchase and read this book. If you would like to contact me online with question or comment please do so at the following email address:

emailto:ganderson61@cox.net

Your comments and reviews are much appreciated!

Other Books in the "*Practical Guides for the Industrial Technician*" Series and available from Amazon!

"*Motion Control Basics*"

"*Industrial Network Basics*"

Made in the USA
San Bernardino, CA
02 June 2015